NATIONAL GEOGRAPHIC

Ladders

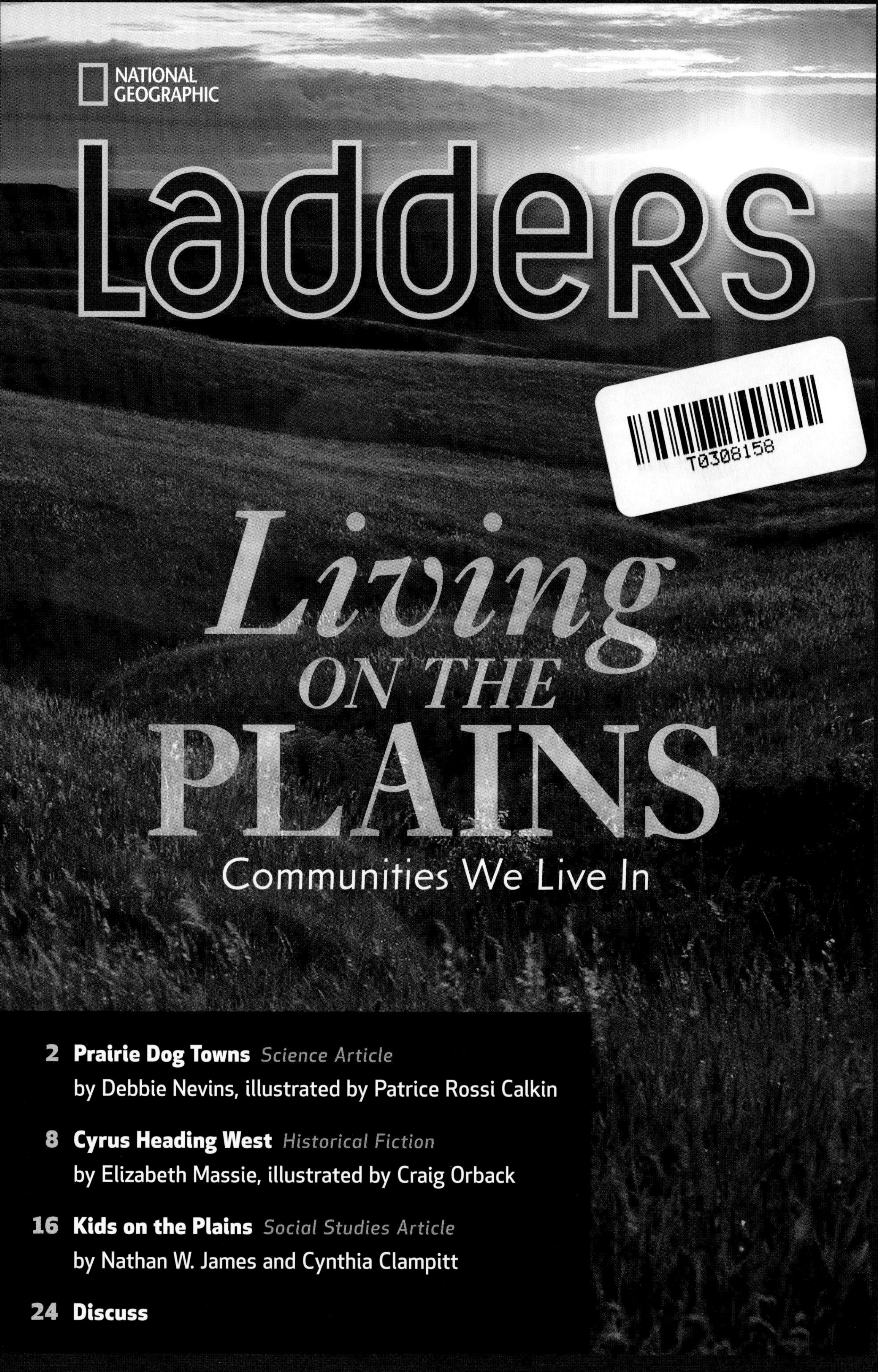

Living ON THE PLAINS

Communities We Live In

GENRE Science Article

Read to find out about prairie dog communities on the plains.

Prairie Dog Towns

by Debbie Nevins
illustration by Patrice Rossi Calkin

As you might have guessed from their name, prairie dogs make their homes on the **prairie**, or grassland. But as you can see, they're not dogs at all. In fact, these cute little animals are **rodents**. Rodents are small animals with big front teeth for chewing.

Black-tailed prairie dogs live on prairies that are part of a region called the Great Plains, a flat, dry, grassy part of the western United States. The Great Plains are the perfect home for prairie dogs because this region has plenty of the sweet, juicy grass they love to eat. They can also tunnel into the rich soil of the plains to build their underground homes. These homes keep them safe from other animals and from bad weather.

Years ago, the Great Plains were full of prairie dog communities. But when people moved onto the plains, they used much of the land for farming and raising animals. Today there's less open land for prairie dog towns, so there are far fewer prairie dogs living on the prairie.

This kind of prairie dog only lives in the state of Utah. It is called a Utah prairie dog.

Home on the Range

Prairie dogs live in large, close-knit family groups—mom, dad, brothers and sisters, cousins, uncles, and aunts. Like humans, they live in large communities called "towns." In its town, a prairie dog's home,

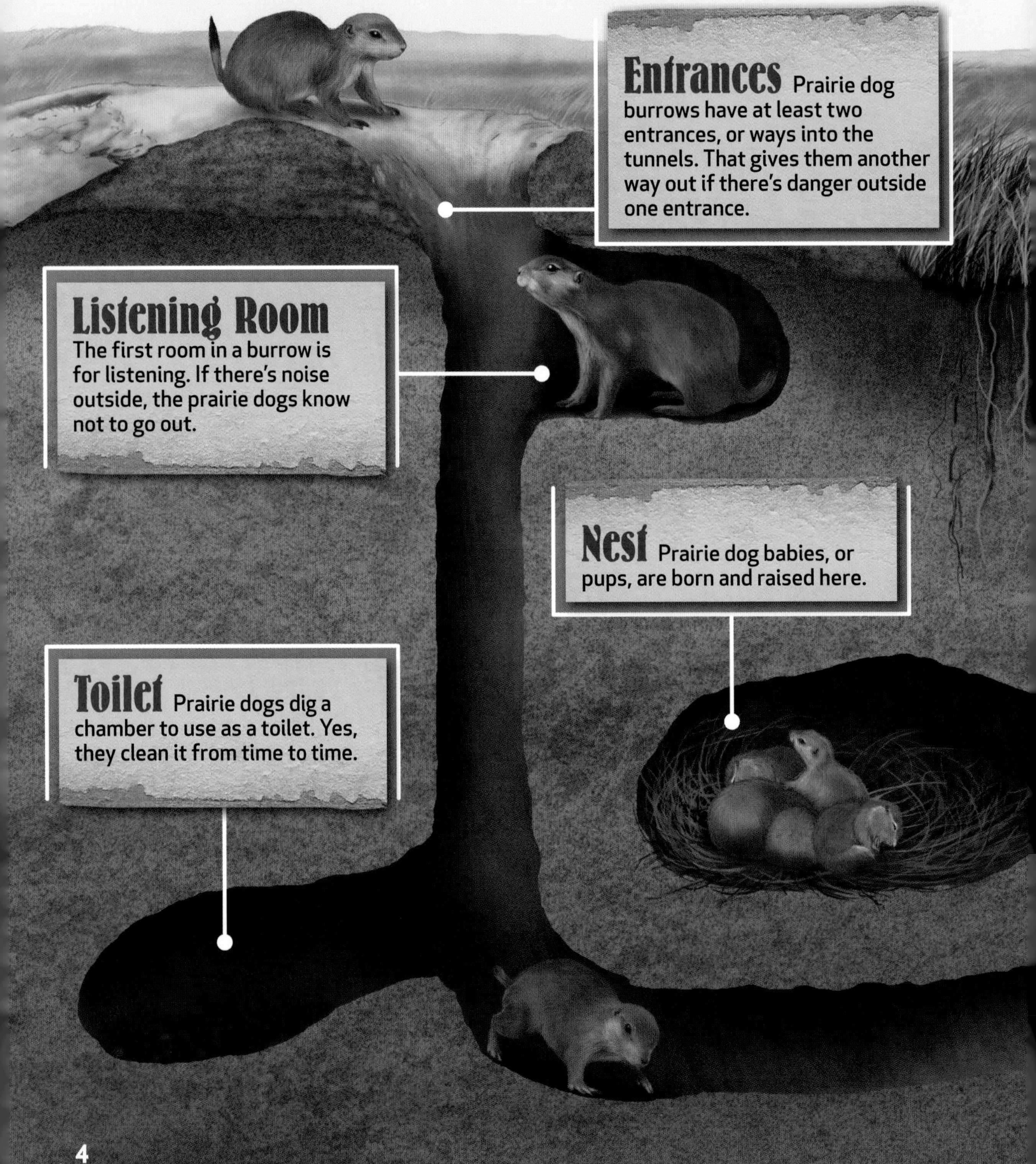

or **burrow**, is like an underground house with winding hallways and several ways to get in. Every burrow is different, but they all have twisty tunnels leading to different **chambers**, or rooms. As you can see from the diagram below, each chamber has a purpose.

A Prairie Dog's Day

Prairie dogs are usually busiest in the daytime. That's when they come out to eat. That's also when they take their pups for a run outside of the burrow. To say hello, two prairie dog friends might touch mouths, or "kiss." They might even clean each other's fur by picking out pesky critters like ticks.

Careful! A coyote, fox, or hawk might come hunting for a tasty prairie dog. Then there's only one thing to do—disappear, fast! Prairie dogs can run really fast even though they're short-legged and chubby.

This mother builds a nest. She makes it before birthing her new pups.

Prairie dogs love to eat a grass called timothy.

These friends greet each other with a "kiss."

Prairie dogs warn others in the community of danger. When the first ones sense trouble, they bark a sharp warning and bob up and down. Then the others cry out. They all dart into their burrows deep in the earth.

Prairie dogs even have different warnings for different enemies. In one cry called the "jump-yip," the prairie dog stands on its rear feet and yips so loud that it often falls over. The yip is believed to tell other prairie dogs that a danger has passed. You can see why being part of a community is the best and safest life for prairie dogs.

Mothers usually check for danger outside before leaving the burrow. Then the pups come out to play if it is safe.

Check In How do prairie dogs keep themselves and their community safe?

GENRE Historical Fiction

Read to find out what life may have been like for a boy and his family moving west in 1870.

Cyrus HEADING

Cyrus, a 9-year-old boy, and his family are moving from Kentucky west across the Great Plains. They are headed to a new, larger farm in Kansas, where land is cheaper. There is also more space to farm and more opportunities to earn money in Kansas. The journey is full of challenges, but together they make it an exciting adventure.

April 24, 1870

Hooray, we're on our way! My brother, Charles, and I walk alongside our wagon instead of riding because the wagon is full of our food and furniture. Our job is

WEST

by Elizabeth Massie
illustrations by Craig Orback

to be on the lookout for big rocks in the rough path we're following. We don't want to break a wagon wheel.

Papa rides a horse ahead of the wagon as my Aunt Millie drives. My giggly cousins, Sissy and Pearl, ride inside, making sure our things aren't getting tossed around while we travel.

May 16, 1870

This is the route we are taking across the plains to get to Kansas. So far, we have traveled west through Kentucky, Indiana, and Illinois. Today, we will be loading our wagon on a ferryboat and crossing the wide Mississippi River to get to Missouri. The ferry has four long paddles sticking out at both sides, which men use to row it across the river.

June 5, 1870

We've just crossed into Kansas, but a terrible windstorm has blown in. We knew it was coming. When Charles climbed a tree this afternoon, he saw a huge, orange cloud on the horizon—that was the storm approaching. An hour later, it hit us.

We're crammed inside the wagon now. It is rocking back and forth. Sand is blowing in the small window we look out of, but we've covered our mouths with pillowcases to keep the sand away. Sissy says she's not scared because our wagon is sturdy, and Pearl reminds us that we aren't far from our new farm. That makes me feel better.

We can't continue our journey until the sand and dust settle down, so I pass the time writing and thinking about how different the land looks here on the flat plains of Kansas compared to Kentucky. In Kentucky, there were forests and mountains near our farm. We had neighbors less than a mile away, on the other side of a hill. In our new state, we might travel a whole day without seeing another family or house. The land in Kansas is dry and flat, and trees are few and far between. All we see is sky and grass. I guess that's how the plains look!

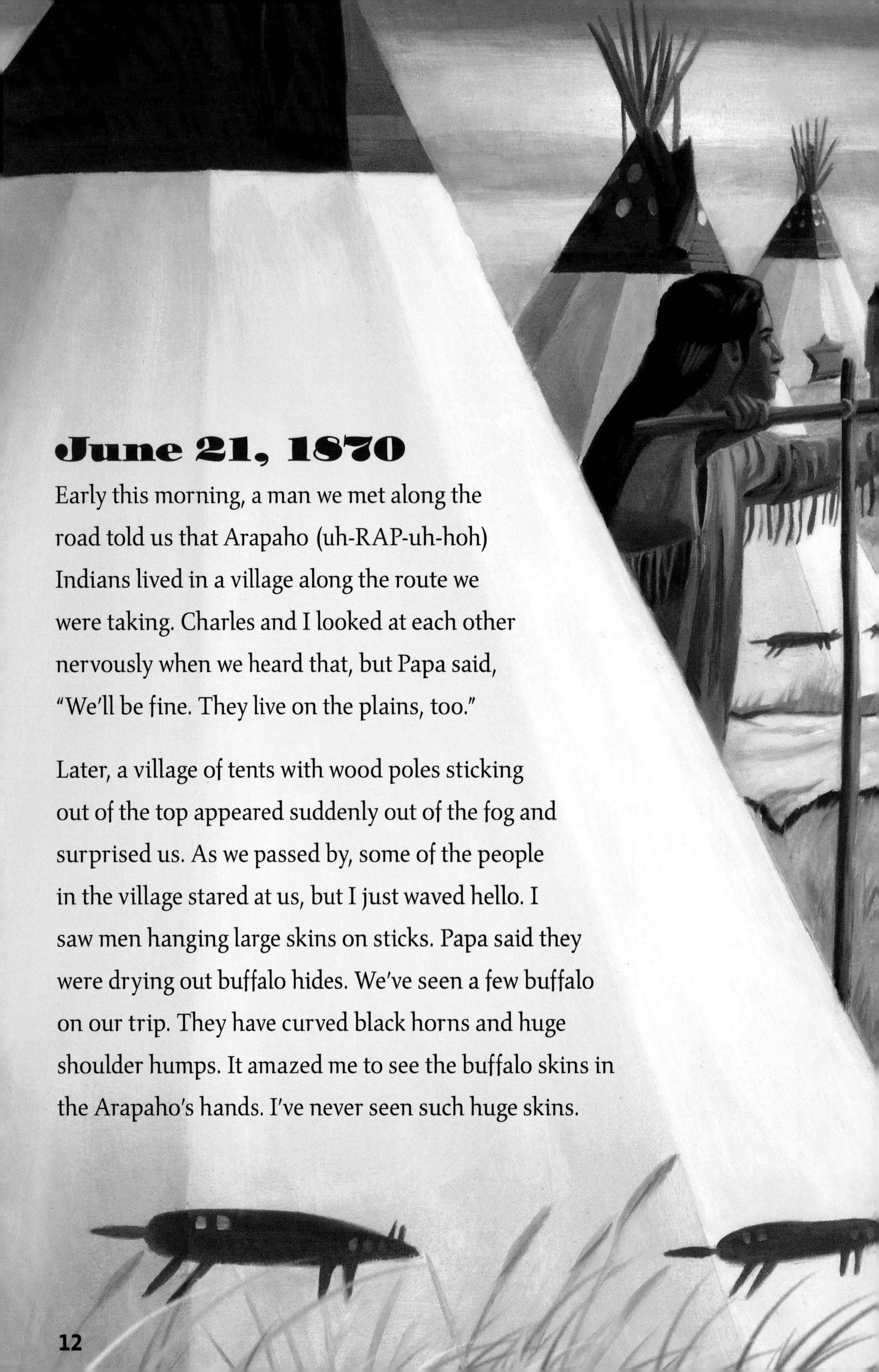

June 21, 1870

Early this morning, a man we met along the road told us that Arapaho (uh-RAP-uh-hoh) Indians lived in a village along the route we were taking. Charles and I looked at each other nervously when we heard that, but Papa said, "We'll be fine. They live on the plains, too."

Later, a village of tents with wood poles sticking out of the top appeared suddenly out of the fog and surprised us. As we passed by, some of the people in the village stared at us, but I just waved hello. I saw men hanging large skins on sticks. Papa said they were drying out buffalo hides. We've seen a few buffalo on our trip. They have curved black horns and huge shoulder humps. It amazed me to see the buffalo skins in the Arapaho's hands. I've never seen such huge skins.

June 26, 1870

We have reached our land! The farm has a small river that flows through it. As soon as Papa said we were home, Charles and I ran to the water and dove in. I felt the cool water on my skin and felt my stiff, dirty clothes soften. It was so nice to take a bath and wash away the dust and sweat from traveling!

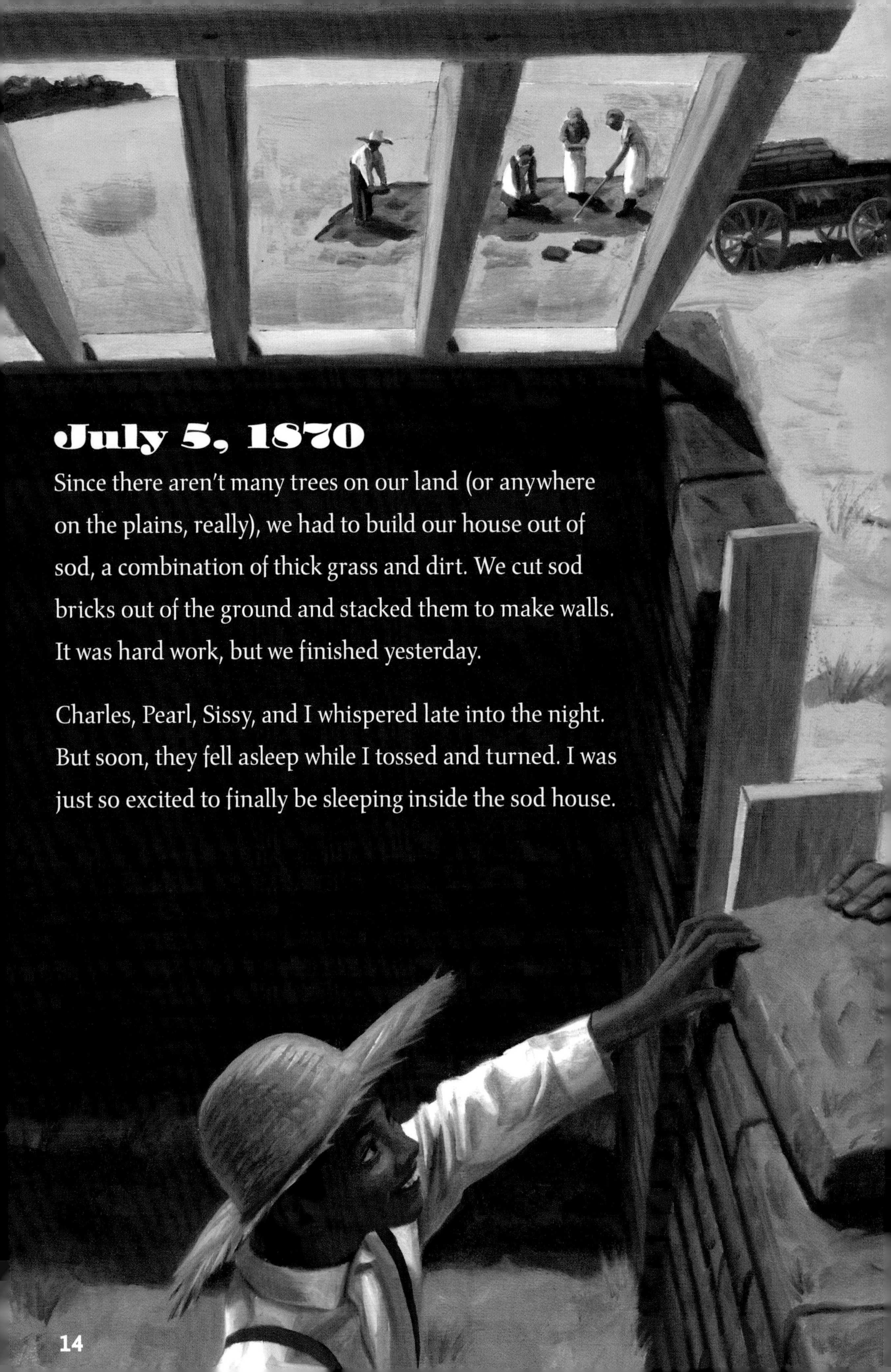

July 5, 1870

Since there aren't many trees on our land (or anywhere on the plains, really), we had to build our house out of sod, a combination of thick grass and dirt. We cut sod bricks out of the ground and stacked them to make walls. It was hard work, but we finished yesterday.

Charles, Pearl, Sissy, and I whispered late into the night. But soon, they fell asleep while I tossed and turned. I was just so excited to finally be sleeping inside the sod house.

Just as I was drifting off to sleep, an earthworm dropped from the sod ceiling onto my face with a plop! I was so surprised, I laughed out loud and woke everyone up. Even with the worm, it was a relief to have four walls and a roof over our heads instead of sleeping inside the wagon.

Today, a boy named Andrew and his father stopped at our farm. Andrew lives on a farm about five miles from ours. We played tag and Andrew said he will lend me some books the next time he visits. I think we'll be good friends. It was nice to have someone new to talk to. I think I'll like it here on the plains of Kansas!

Check In What new things did Cyrus see when his family reached the plains of Kansas?

GENRE Social Studies Article

Read to find out about life on the plains of the United States and Argentina.

Kids on the Plains

by Nathan W. James and Cynthia Clampitt

Many countries around the world have plains. The plains of the United States are called the Great Plains. In the country of Argentina, the plains are called the Pampas. Countries might have different names for their plains, but all plains are flat with very few trees. Let's take a look at the plains in two different countries.

> The Great Plains are known for their long stretches of flat land. They also have good farming soil.

That's me

Hi!

My name is Amanda. I live on a farm in Nebraska. Nebraska is a state in the middle of the American Great Plains. When I look out of my bedroom window, I see fields of green corn and golden wheat blowing in the wind. My family's farm is surrounded by many other corn and wheat farms. That's why the Great Plains are called "America's bread basket," because corn and wheat are grains that are used to make many foods, including bread.

Corn, my favorite!

Corn and wheat aren't the only foods that are grown on the plains. We're having a family picnic today. Most of the food we'll be eating will come from around the Great Plains. The beef for our hamburgers comes from our neighbor's cattle ranch. The buns are made from local wheat. We even grew the lettuce, carrots, and tomatoes for our salad in a garden beside our house.

Wind-Blown

Fields aren't all that I can see outside my window. The flat land of the Great Plains is also dotted with **wind turbines**, or machines that have blades that spin in the wind to make electricity. They look like huge metal flowers with petals that turn in the wind. The turbines on our farmland give power to our town and several of the nearby towns. Nebraska is not close to any mountains that could block the wind. So the flat, windy plains are perfect for twirling turbines.

The tower of a wind turbine can be as tall as a 30-story building. The blades can be as big as the wings of a large jet.

Hola (Oh-lah)!
Hello!

My name is Mateo, and I live in the "great plains" of Argentina, which is a country in South America. We call our plains the Pampas, but I think they're a lot like the Great Plains in the United States. The Pampas are flat and grassy. We grow a lot of food here on farms. The fields are also the perfect place for cattle to spread out and graze, or eat grass.

The Pampas are flat and grassy, but they're just a short distance away from the mountains and the ocean. In fact, I just went hiking in the mountains with my mom, and I have gone fishing on our cousin's boat many times. It's really fun!

You can walk for miles and miles on the Pampas before finding shade from a tree.

I ride horses in these hills, called "foothills." They are at the base of the Andes Mountains.

My Dad, the Gaucho

When I'm not hiking, I practice my horseback riding. I live on an **estancia** (ay-STAN-see-ah), or ranch. My family has lots of horses and cattle. We raise the cattle to sell at the market. My dad is a **gaucho** (GOW-choh), which is a lot like an American cowboy. He rides a horse to **corral**, or bring together, herds of cattle. Sometimes the cattle wander all over our land.

Gauchos like my dad dress differently than the cowboys you have probably seen. They wear a wide belt to hold their tools. They also tuck their pants into tall leather boots. The winds are always blowing across the Pampas, so when the weather turns cooler, the gauchos wear a serape (suh-RAH-pay), a kind of wool cape, or coat. A gaucho is never without his bola. Bolas are leather cords with three iron balls at the end. Gauchos use them to help round up cattle. My dad can throw the bola so it wraps around a cow's legs to stop it from running away from the herd.

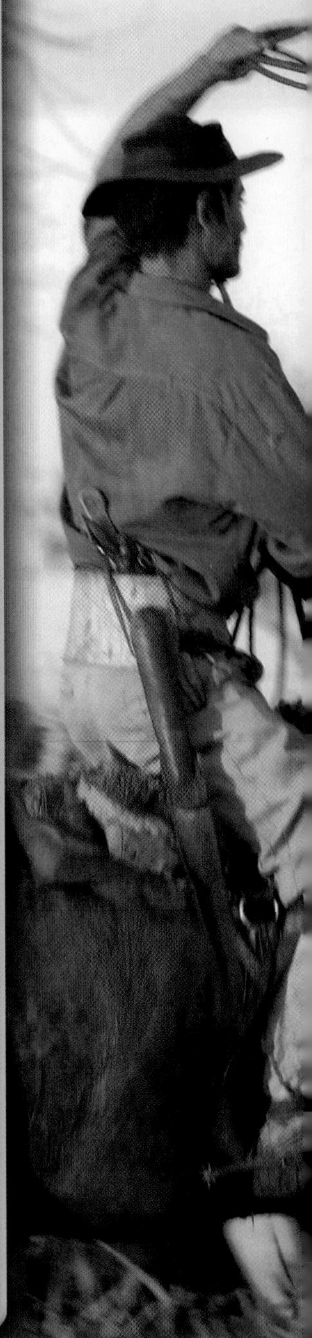

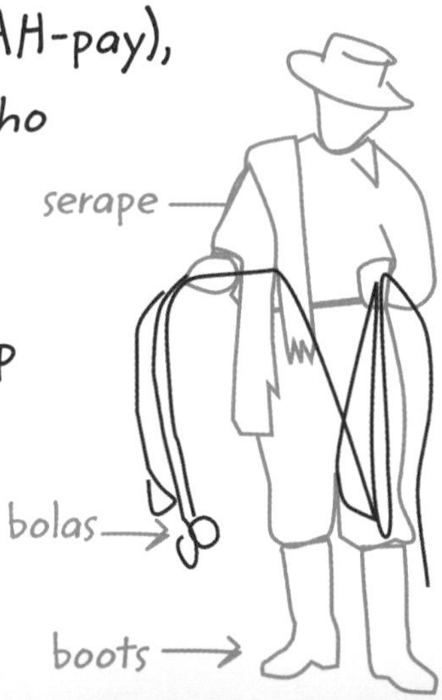

The rich, flat land of most of the plains in the world is good for growing food and raising cattle. No matter where a plains community is located, the land shapes the lives of the people living there.

Gauchos also use a long rope to herd cattle. They loop the end of the rope around the cow's neck. Then they guide it to safety.

Check In Compare the American Great Plains and the Pampas. How are they alike?

Discuss

1. What connections can you make among the three selections in this book? In what ways are they related?
2. How are prairie dog communities like human communities? How are they different?
3. Where did Cyrus and his family begin their journey? Where did their journey end? Compare their old farm and the land around it to their new farm and its land.
4. How does the wind play a part in life on the plains?
5. What do you still wonder about life in a plains community? How can you learn more?